Infinite Reg

Scott Axelrod

Infinite Regress.

Printed in the United States of America.

Published by Tandeta Books (Newton, MA).
Library of Congress Control Number: 2022931111.
FIRST EDITION
February 2022.

ISBN 978-0-9998455-8-5

In gratitude for all people, who have struggled to balance creative outbursts with disciplined follow through.

This little manuscript got started as journal entries intended to lead to writing of the first column in a regular series, "Infinite Regress", to appear in the inaugural issues of the journal "Tandeta", entitled "Epiphany", published by Tandeta Books[E1]. As it grew in size, the decision was made to turn the journal entries into this standalone manuscript and to place in the "Epiphany" issue a short piece "Infinite Placeholder Regress"[E2] that refers to this manuscript.

To maintain the authenticity of a journal, the body of this manuscript is a lightly edited version of the original journal entries.

Caveat Emptor

Infinite Regress

January 2022

Publisher's Note

On the evening of the day on which Scott Axelrod and I decided to publish *Infinite Regress* as a sibling publication to the *Tandeta* journal, I shared with a different male friend of mine Ludwig Wittgenstein's famous remark that mathematicians *do not come to blows.* This is the case, my interlocutor and I agreed, because mathematicians converse according to strict rules they agree to consider productive of truth (or what passes for it). This means that in order to revolutionize mathematics—or any other system of thought—one must disrupt the symbolic order (for example, as Gödel did by laying bare the inevitable disjunction between any system's powers of expression and its powers of proof). As Wittgenstein also suggested, instead of following rules that say: *when this symbol appears you must next write that one*, you should follow a rule that says, "you write what you write!" Scott Axelrod's writing helps to create the conditions for such disruption and intellectual revolution by documenting the operations of thought that constitute such a symbolic order for one individual—himself.

In so doing, he follows a long line of philosophers who have enjoined us to examine our own minds as the prerequisite for any serious intellectual exploration of the world—whether that world is conceived of as "outside of" or, as in Western phenomenology and Eastern religion, deeply imbricated with the self. From Socrates' "know thyself" to Wilfred Bion's insistence that a capacity for psychoanalytic insight must be grounded first and foremost in the analyst's own self-observation, great thinkers have stressed the value of thought taking itself as its first object. In his writing—which first captivated me in his book, Off the Deep End: Diary of a Mathematician—Scott interweaves reportage and reflection on ordinary life with musings on philosophy, psychology, spirituality, physics, mathematics and politics. He calls the result "thought-swirl" and claims to see in it hints of a "madman" side. But I see it as eminently sane and to the extent humans can be, logical; I view it as an illustration (psychoanalysts might say an enactment) of what Alfred North Whitehead, in his 1929 The Function of Reason, calls the nature of conscious experience.

At any moment of our lives, Whitehead writes, "there is a focus of attention, a few items in clarity of awareness, but interconnected vaguely and yet insistently with other items in dim apprehension." Moreover, the moments of experience differ with our moods, situations, and physical states: "our variety of phases," Whitehead tells us, "is infinite." In this first issue of *Infinite Regress*, Scott shows us a finite sampling of his phases by recording the movements of his thoughts at different times and locations, in writing sessions of various durations, bounded or marked by interpersonal reactions of various kinds. The effect is of a mind circling around and through the many impressions received in an embodied, societal, and intellectual life, approaching again and again key scientific and mathematical concerns. The approach and re-approach occurs not, however, in a merely repetitive way, but as a progressive opening up of new avenues of inquiry and reflection on some of the most fundamental ontological puzzles of our time.

Finally, to support my judgment of the value of this project, I call again on Whitehead. Marveling in the The Function of Reason that the "extreme abstractness" of mathematical physics has "proved to be of such importance," he notes that common-sense often dismisses as "[a] tenuous, unpractical, waste of time," speculations that later turn out to have very concrete value. In fact, he writes, "[a]bstract speculation has been the salvation of the world—speculations which made systems and then transcended them, speculations which ventured to the furthest limit of abstraction. To set limits to speculation is treason to the future."

So publishing *Infinite Regress* is, perhaps, one of my own small acts of faith in our collective future.

Kimberly Gladman

Newton, Massachusetts

January 2022

Contents

Preface

Modified from 2021-12-15 journal entry.

The image motivating the name of this column is an infinite reflection mirror[1], in which mirrors reflect the reflections of other mirrors, *ad infinitum.*

The statement "everything you see in an infinite reflection mirror is a reflection" seems to lead to a vicious infinite regression[2].

Not everything can be a reflection of a reflection. Doesn't there have to be some core element of reality that gives rise to the primordial reflection? Besides the question of input, what of value gets output by a true infinite reflection mirror?

I chose the name "infinite regress", rather than "infinite reflection", for this column as a reminder to myself, and warning to you, of the dangers of being sucked into too many layers of reflections on reflections.

My plan for this column is to invite you into my head to see the ungainly process of my reflecting on matters of concern to me, like the nature of reality. In order to fairly reflect my thinking process, I will try to keep to the informality of a personal journal, written by an author who sometimes has a reader in mind, but often is just caught up in their own thinking. In fact, this kickoff for what is planned to be a monthly column consists of a lightly edited version of my personal journal entries that I made when thinking about what to write for this

[1] Credit to my brother David "Daxe" Axelrod for bringing the metaphorical potential of infinite reflection mirrors to my attention.

[2] See: "Infinite Regress - Virtue of Vice?"[E1] by Anna-Sofia Maurin in Hommage à Wlodek. Philosophical Papers Dedicated to Wlodek Rabinowicz. Ed. T. Ronnow-Rasmussen, B. Petersson, J. Josefsson & D. Egonsson, 2007.[E2].

column. The dates of the original journal entries are indicated in the text.

"If you want to follow me"[3], I will provide the original handwritten diary files that this essay is based on, along with the audio recordings I made of them. You can also read my memoir Off the Deep End: Diary of a Mathematician[E4], which for present purposes you can view as practice documenting my trips into madness and/or the infinite.

[3]See "We're Not Gonna Take It" by the Who[E3].

Part I: Rambling

Modified from 2021-12-12 journal entry.

First off, I have to tell you, dear reader, that I am going to keep some things private from you. I had a journal entry (on Dec. 8) about a conversation I had with Denise (my wife) the other day, but I won't include that for distribution, at least not for some time. Going on the recent theme in my journal that I am the author – the God – of my own imagination, I am declaring that my self-God has many sides that will not be all revealed.

I want to get a take on people (like me) who have had a manic sense of being a special parts of God's plan, or who have had the "unitive experience" (that we are all connected as "one" and one with God.) As subject of this "take", I will also include people who (again, like me) have felt the reality of the solipsistic interpretation of quantum mechanics, which resolves quantum mechanical mysteries by saying that I can only "assign reality" to what is accessible in my experience and memory at one instant in time. Perhaps this is not so limiting, since the here and now of personal experience is all you or I or Descartes can actually know anyway.

As I explained in my last paper written for college, if we restrict quantum mechanics to explaining what is in one observer's "head"[1] at any one time, which may include observations of other observer's observations, then the standard rules of quantum mechanics gives a complete and consistent explanation of that observer and the world that they perceive. If we restrict to explanations of the memory of results of

[1] I use the phrase observer's "head" here because, ultimately, we would like to understand conscious observers like ourselves. However, in actual physical calculation, we could be talking about an individual particle as an "observer".

interactions, which may bring together many observations of various observers, then we can make sense of the fact that quantum mechanics only makes probabilistic predictions and the fact that results of different observers are "quantum" entangled.

I could blather on comparing the micro-God of my own solipsistic world with the macro-God of the many worlds multiverse. I could develop a full theory of consciousness and free will from either perspective. Or I could talk about how solipsism and the multiverse are extreme points in a range of world views, whereas a more centrist and practical view of the mechanisms of interactions of human beings belongs to the realm of ordinary science, economics, psychology, sociology, politics, and the saner sides of religion and philosophy.

Here is a key question for me:

> What should a well – but perhaps unevenly – educated person do in the modern world when they have had their fill of the tameness, but have no guidance about how to venture into the untamed?

A person such as one in the above question may have had their own idiosyncratic glimpses of the wild side (of reality), but they realize that (assuming anyone else actually exists) there are many people more experienced at exploring the wild side. The "sex, drugs, and rock roll", ecstatic bliss, revelation, enlightenment, and hyper-creative communities, all with various mixtures of reality and fantasy, are so well developed that they appear to be tamed to the point of being like guilds with bars to entry. It's no surprise that what might start out as individual, wild, crazy rides through reality become captured into definable life-style choices.

Back in 1961, Paul-Michel Foucault[E1] wrote in Madness and Civilization: A History of Insanity in the Age of Reason[E2]:

> The language of psychiatry, which is a monologue by reason *about* madness, could only have come into existence in such a silence [in which dialogue with the madman has been shut off].

I'm inconsistent. On the one hand: I am trying to reclaim the right to be in dialogue with my madman side; and, when I come into contact with that side in others, I am fascinated and drawn in. On the other hand: I remain wary and maintain my own "monologue by reason *about* madness".

Is all my intellectualizing just an elaborate set of mirrors to reflect on my and other's madness from a safe distance?

I don't think it is a coincidence that my mind wandered into the above little discussion of madness as preface to my turning to think about a piece I might write as the first piece for a monthly column called "infinite regress" in the inaugural issue of the journal Tandeta, which will be on the theme of "epiphany".

Kimberly Gladman is hoping to get the first issue of Tandeta out by the end of January. Tandeta Books[E3] is Kimberly's "publishing imprint for good books, simply produced." The Tandeta web site explains that:

> "Tandeta" is a word Polish artist Bruno Schulz used to mean treasures found in unexpected places.

Gee, it would be nice to write an unexpected treasure about epiphany. But to cook something up I'm going to have to mix up and break through the hardened neurotic batter enclosing and protecting the gooey madness at the center of my brain.

The crusty-neurotic-batter-surrounding-gooey-madness metaphor above is a distorted reflection of thoughts I have had about genius being a mixture of $\overline{\text{creativity+discipline}}$. In this distortion, somehow "discipline" has been distorted to "neurotic" and "creativity" has become "madness".

You have now entered a world where I am colliding metaphors to think about my thinking. I have given a hint that there is at least one profound philosophical movement (with Foucault as protagonist) and some deep science (quantum mechanics) in the mix somewhere in my thought swirl. You have to decide if it's worth your time to see if I have the diligence to select good ingredients and the talent to actually cook up a nice dish with those ingredients. Do you really want to put up with digressions like the one in this sentence pointing out that talent + work is a reflection of the psychosis + neurosis reflection of creativity + discipline?

Do you really want to hang out with someone trying to bootstrap their own brain, especially if – truth be told – it is just another effort at rebooting after many crashes? There is crap (oops, I should say "there are tasty ingredients") on the hard disk, but the updates are all in conflict. I don't even have a stable operating system, no less the ability to run a program that will allow me to push a pair of computer-related phrases on top of the pairs talent + work, creativity + discipline, and psychosis + neurosis.

OK, just to give you a chance at figuring out what the hell I'm talking about and to provide another opportunity for you to bail out of reading this twisted journal entry / essay, I'll give you a provisional stack entry as a placeholder for the entry I just said that I am unable to produce. Here it is: (program) runtime + (operating system) administration .

Just a reminder: You might want to bail out right now. Or at least keep up a skeptical shield.

The book title I hate You – Don't Leave Me: Understanding the Borderline Personality by Jerold J. Kreisman[E4]" just popped into my head. This is one of the many barely read books in my collection. The dough of my neurosis, which I'm kneading with you currently, is a bit different than the borderline personality batter. The book subtitle summarizing my neurosis is more like "I love you – but maybe you're better off going - but secretly I am hoping you are hoping it will be worth staying just a bit longer to see if the system reboots, the recipe gets cooked, this essay comes to a coherent conclusion, or we come to some grand epiphany about epiphanies.

My mother characterized my father's schizophrenic mind as having very many "loose firings". I'm just waking my brain up in this essay by putting some loose firings on the table (or in the batter).

I wish I could quickly think through many grand ideas, pick out the most fertile seed, expertly incubate the seed on surrounding nourishing fruit, and plant the new idea tree in a field to start a thriving orchard. That's the way my Ph.D. advisor often worked. Name drop: He's Edward Witten. He was so adept at distinguishing fertile seeds, it was hard to respond when he commented on my seeds of ideas by saying, "I think we can do something better with our time." There was no room for my loose firings.

I could follow the output of Witten's metaphorical process,

$$\boxed{\frac{\text{fertile seed + nourishing fruit}}{} \text{ become tree in new orchard}},$$

but I couldn't get Witten's process to run on my hardware. My hardware is more designed around the following process:

$$\boxed{\begin{array}{l} \dfrac{\text{loose firings / creative hunch /}}{\text{interesting ingredients / grandiose fantasy}} \\ + \\ \dfrac{\text{reflection / discipline / cooking / reality constraints}}{} \\ \text{become unfinished project.} \end{array}}$$

[2:45 pm : Stop to take Dari (my child) to see a friend and go to Staples, 3hr, 43 min session]

Part II: Focus

Modified from beginning of 2021-12-13 journal entry.

OK, so let me try to gather together thoughts more specifically focused on writing something for the "Infinite Regress" column in the inaugural issue of the journal Tandeta, which is slated to be be focused on the topic of "epiphany".

But first, a quick digression: The title of yesterday's journal entry and the current title combine to give rambling + focus. This combination can be pushed on the stack of phrase pairs that I was building yesterday. With some editing, the stack (from bottom up) currently is as follows.

Creative Burst + Follow Through Stack (currently)

1. creativity + discipline
2. (psychosis / madness) + (neurosis / control)
3. talent + work
4. (operating system / executive functioning) + (program runtime / flow)
5. (fertile seed) + (nourishing fruit)
6. (loose firings / creative hunch / interesting ingredients / grandiose fantasy) +
   (reflection / discipline / cooking / reality constraints)
7. rambling + focus

Why have I listed various ways of naming a pairing of a creative burst with a detailed follow through? It's because I'm hoping for an epiphany that will help me find a good balance between the sides of each pair and to get perspective on the whole stack.

My brother Daxe has a saying, "You get paid to produce and you pay to create." I'll add this to our stack[1] as entry: 8. create + produce.

Daxe's saying is a take on the universal struggle between self-fulfillment, providing to society, and receiving from society. Implicit in this is a long riff that Daxe can (and has) given as a creative economist on choices, tradeoffs, sustainability, and time horizon.

It's interesting what I've done here in bringing up Daxe. I'm falling into the pattern of our discussions since childhood. Daxe would say some seemingly (to me) wild ideas and I would work to make sense of them, taming them into something that can be better (again, in my opinion) harnessed and communicated as something sensible and reasonable.

We have stumbled upon a new stack entry:

9. (wild idea) + (sense / reason).

Let me let my inner child talk for a moment:

> *Why do I always have to be the good boy?*
>
> *I can do, I have done, the wild and crazy thing too.*
>
> *When other people go on a wild and crazy ride on the river of reality, I'm willing to hop on the boat and keep it from tipping over, while I add steering and stability mechanisms.*

[1] For the reader interested in the full, twisted experience, I will mention that the entry number 8 on the stack entry above is printed using the LaTeX counter `stacksize`.

> *I try to fit onto their boats (play their games) where they are the captain (their rules), but I don't know how to get them to let me navigate. And I don't really feel I have a right to. After all, it is their boat and their wild ride.*

I feel like a ship builder that never gets others to ride on my ship with me[2].

In an interview with Science Lives[E1]. the great mathematician Isadore Singer (who brought me to MIT) said:

> *"I am always exploring" is a good metaphor.*
> *And I think of the vast forest.*
> *And I think, if I can only just start a path in such and such a direction, I am quite happy.*
> *And when it turns out that that path turns into a highway, that's really remarkable, but that's the past, we're off to looking for new paths.*

This bring to mind Isaac Newton's famous quote[E2]:

> *I do not know what I may appear to the world, but to myself I seem to have been only like a boy playing on the seashore, and diverting myself in now and then finding a smoother pebble or a prettier shell than ordinary, whilst the great ocean of truth lay all undiscovered before me.*

I guess the truly great ones keep their sanity by having a perspective with some humility. Gaining perspective is an admirable accomplishment in itself, since they have collectively lead the change of the intellectual (and physical) landscape of humanity from a few individuals playing at the beach, to large crews on globe-encompassing ocean liners, to voyages into space.

---

[2]By the way, a ship is a large seagoing vessel, which you could put a boat – a small vessel – in. The Columbo episode that I watched last night to help me get back to sleep after multiple wakings was called "Troubled Waters". It's very entertaining how Columbo is confused by the difference between a boat and a ship throughout the episode.

Everyone on all those voyages, as well as those that stay behind, strikes their own balance between the fulfillment of exploration, workmanship of helping to build the ships, and the comforts of staying snuggled at home.

Oh my, how loud and insistent are some voices saying, "Play my way".

Who is preparing room on the space ships for the quiet, thoughtful children?

And who is preparing room for those who are preparing room for others?

After only finite reflection on the questions above, I come to new questions:

> What is the role of the General Organization Director (GOD)?
>
> If we are all GODs of our own minds,
> and a reflection of the GODs in other minds,
> What is the right way to be?

Part III: Try Again

Modified from end of 2021-12-13 journal entry.

Dear reader, I've given you a questionable taste of my mind. I tried to focus, after rambling, but you just witnessed how I quickly orbited out to spaceships and talk of God, ending, essentially, with the desperate questions, "How should I be?"

Please, let me know if you have a better answer to the question "Who am I?" than I do. But, in the meantime, I'll try to refocus on Kimberly's ideas for the Epiphany issue.

Kimberly sent me drafts of two or her essays to give me an idea of context for the issue. The essays were "Aftershock" and "Quantslut: A laywoman's musing on the 'hard' sciences and math." In "Aftershock", she talks of the "phenomenon ... of a retrospective realization that a process directed toward a particular end has been unfolding for some time." In "Quantslut," she talks of several theorems and papers that she finds inspiring. The subtext is that perhaps a culture of revealing tandeta (treasures in unexpected places) that she is trying to start will lead people to have epiphanies that will lead to groundbreaking (or should I say "shipbuilding") progress.

Upon multiple-reflection, I understand that Kimberly putting these things in front of me, combined with the crucial act of inviting me to participate, quickly led me to a fully disassociated response.

The frustrated little boy in me wants to say:

> *But, but, but ... you can't just throw out your raw musings and expect someone else to have the epiphany and do the hard science follow through and do it on your boat.*

On the other hand, the recovering (but not fully recovered) grandiose, neurotic, scientist-mathematician in me wants to say:

> *Damn, you are a sexy muse Kimberly. I realize just now that my entire life has been leading to this point where I can receive the tasty ingredients and creative spark you have provided to light my fire to cook up an amazing new vessel*
> – What the hell, I'm ranting now, are you really going to expect me not to mix my metaphors –
> *to probe the very nature of reality more deeply than ever before!*

As usual, my neurotic little boy needed to be assuaged first, before I could let grandiose dreamer guy run free. So I have subjected you, dear reader, to Parts I and II, and this start of Part III, wherein I intellectualize a whine about not knowing how to play.

But now I'll let my internal Grandiose Ontology Dreamer (GOD) speak.

The solipsistic interpretation of quantum mechanics is fine for seeing the little picture. But if we really want to contend with the big picture, we need to deal with the possibility that there are "many worlds" associated with varying results of quantum measurements that have undergone "infinite reflection" by multiple observers, i.e. an infinite process of observers observing – and becoming entangled with – each other.

The world is waiting for the next step beyond the "shut up and calculate" school of physics. We need to combine deep insights and answer unquestionable questions[1] about the nature of experience within the context of a well-defined mathematical model. And it needs to lead to results that are testable and get verified. This will lead to a profound advance in our understanding of existence itself.

[1] The phrase "unquestionable questions" alludes to the title of the book Speakable and Unspeakable in Quantum Mechanics by J.S. Bell[E1].

All of this, and the essay that Kimberly gave a copy of to me as context for the epiphany issue ["Links Between Physics and Set Theory" by Bruno W. Augensten, Chaos, Solitons, and Fractals, Vol 7, Issue 11, Nov. 1996, pp. 1761-1798] ^{E2} should be documented as providing the key spark to unravel the whole mystery.

I will now sketch how all of this will unfold. This will be a down payment to secure funding, love, and accolades. The sketch will allow me to fill in the details in the future.

[2:30 pm : Lunch Break with Denise after 2hr, 51 min session]

Part IV: Explanation

Dedicated to my father, who would have been 83 today.

Modified from 2021-12-14 journal entry.

Fortunately, or unfortunately, I have had some time to think after writing the "try again" entry yesterday. This means that my mind has fired through a lot of loose thoughts, all motivated by figuring out what to say in this "explanation" essay.

When I stopped writing yesterday, I thought the plan was to string together various thoughts, which I have had over many years, about an "ultimate theory".

I was going to start talking about: How the article "Links Between Physics and Set Theory" (which, yesterday, I had declared would be "documented as providing the key spark") got me thinking about the axiom of choice and quantum mechanics. And then how that brought me to non-separable Hilbert Spaces (which, for those in the know, don't have a countable L^2 basis). This brought me to super-selection sectors in quantum field theory, which are basically parts of a quantum field theory that don't talk to each other.

I did a little calculation to verify that there is no overlap[1] between sectors built upon different ground states of a quantum field theory consisting of only a Higgs Boson[2]. Actually, this calculation was one I should have done (and perhaps did do?) a long time ago as warmup to a project I had in mind which would have shown that an integra-

[1] no overlap: the inner product is zero

[2] Each ground state of the pure Higgs quantum field theory (on the spacetime of special relativity) is a vacuum built around a solution to the classical field equation whose derivatives go to zero at infinity.

tion over the moduli space of classical solutions gives good results[3] for compact[4] target spaces[5]; whereas, for non-compact target spaces, different (behavior near infinity of) solutions correspond to different super-selection sectors.

The intent of the project proposed above is to build tools to provide a full mathematical framework for the Everett many worlds interpretation of quantum mechanics. The idea is that each world would correspond to a different super-selection sector. As time (in an appropriately generalized sense) progresses, super-selection sectors proliferate. Specifically, the splitting into different sectors would be a precise rendering of Everett's notion of a measurement observation causing a split into many worlds, one for each outcome.

The essential new component I propose is that the world (Hilbert Space) only splits (into super-selection sectors that don't overlap with each other) once an observation has become correlated with an effectively infinite number of observations – which we can call infinite reflection!

Even a child can point out, as my seventeen-year-old child Dari did, that the Higgs fields example isn't *really* a good example because the infinite copying of a single observation, which in this case corresponds to the duplication of the symmetry breaking value of the fields to all of space, involves infinite distances, and so (according to relativity) won't be communicable in a finite amount of time. Dari was, of course, correct since this example involves very long distance ("infrared") phe-

[3]For three-dimensional Chern–Simons quantum field theory defined for a compact manifold without boundary, I had proved that the perturbation series defines manifold invariants when one integrates properly over the moduli space of classical solutions. (Technically, I only handle the case when the moduli space is free from singularities.) I gave several talks on this, but only partially published the results.

[4]A manifold is *compact* if any collection of open sets that covers the manifold (i.e. whose union is the whole manifold) has a finite sub-collection that covers the manifold.

[5]I use the term "target space" above since it emphasizes the known relationship of three-dimensional Chern–Simons quantum field theory to Chern–Simons string field theory. Also, the term "target space" is more general than the term "spacetime", which would falsely imply that the target space will be equipped with a notion of distance of the type in general relativity (an indefinite signature *semi-Riemannian metric*).

nomenon; however, the same idea can apply to short distance ("ultraviolet") effects. Since (non-topological) quantum field theories allow for an infinite number of physical modes interacting with each other in an arbitrarily small amount of space, I think there is room in a finite space-time to see the separation of information phenomenon.

An analogue of how quantum worlds can be separated is how (classical) black holes separate information into inside (the "null horizon") and outside pieces. Many people are now playing with this kind of idea. At some point it was all the rage in parts of the physics community to think that understanding black holes better is the key to understanding how to merge quantum mechanics and general relativity, see ["Black Holes: the Harmonic Oscillators of the 21st Century", talk by Andy Strominger in Perimeter Institute Colloquium, March 2014][E1]. I'm suggesting that the separation of information effects in quantum mechanics and black holes are related, and both have to do with processes of "infinite reflection".

- Dear Reader, I've sort of lured you into listening to some physics babble.
- In hindsight, I see that I have tricked you and myself with the lead-in words: "What I was going to say was."
- So, if you haven't left or skipped ahead, you have no doubt been waiting for me to get to the new thing I decided to say instead of the thing I decided not to say but have somehow gone on about for a few pages anyway.
- OK, are you good and weary-ready?
- The "new" thing is more psycho-social stage setting rambling.
- Oh goodie.
- Wasn't I supposed to have finally gotten enough of that out of my system in the past days so that I could get to the "meat" of the explanation today?
- Unfortunately, we are discovering that I am capable of an almost infinite amount of scattered reflections, scattering off of each other in an arbitrarily small amount of conceptual space.

- I hope you are capable of an almost infinite ability to shield yourself from irrelevant information (to morph the technical term in quantum field theory, where infinities are avoided by "shielding").
- But, in my heart, this new flood of psycho-social rambling contains highly relavant insights which are key to my being able to lead you out of this black hole of dense, stuck thought that we are (OK, I am) trapped in.

The key words in the previous sentence, which I sort of semi-consciously placed there as guide post to get out of this black hole of the mind, rather than gravitate towards all the heavy insights that lured me in[6], are "rambling" and "lead"

- Think of the black hole part of my mind as the sub-conscious, or pre-conscious, or semi-conscious part filled with vague awareness of gestating embryos of ideas, thoughts, and feelings.
- Let's not get distracted, at least not right now, by worrying about the rights of unborn ideas.
- Let's shoot for our first guide-post word, "rambling".
- Now that I am trying to bring that word into conscious focus, I recall that I wrote (on page 9 , at the beginning of the section "Focus") that rambling+focus was a phrase that could be pushed on the "creative burst + follow through" stack.

Note to signal part of my brain that it has been noted so it can quiet down:

- All of the "creative burst+follow through" pairs should have a place in our dreamt of big theory.
- They should relate to one another in a manner somehow generalizing Bohr's complementarity principle in quantum mechanics and philosophy.

[6] Get it?
Gravitating toward heavy things.
Lured in, like a black hole.
Ha. Ha. Ha.

- My mother, who was the executive director of NAMI NJ, the New Jersey Alliance for the Mentally Ill, and I have often shared "rant letters".
- Those are the letters you have to write to get stuff out of your system before you can sit down, focus, and write a constructive letter.
- The rant letters are the inner child running wild before the adult hunkers down to work.

My inner child became a bit neurotic trying to shield the madman hidden inside. That's the "frustrated little boy" yesterday, on page 13, who said, *"But, but, but ... you can't just throw out your raw musings ..."*.

My breakthrough *yesterday* was to permit the inner madman to say (on page 14):

I can ... probe the very nature of reality more deeply than ever before!

And my breakthrough *today* is to realize that I have made an advance by transitioning from naming the act of these raw (child and madman) voices as "ranting" to respectfully calling the acts "rambling".

Both voice have a right to come out. They both need to come out if I am going to live fully. These voices are key to enabling me to follow the advice that Edward Witten once gave me, which was:

You need to learn to work when you don't know what's going on.

Here is another comment that Edward once said to me:

You should be a leader in the field.

But my boy/madman hadn't learned how to play with the big boys (and "big boys" they often were). I had no idea how to lead them to play my little boy / madman games.

I learned that I could play the popular big boy games fairly well, but something inside me was dying. I literally cried, "I just want to take my ball and go home."

The responsible adult me separated from the inner boy / madman infinitely reflecting inside the "shelter of my mind"[7].

My big epiphany today is that the civilized ego-self that I have constructed needs to stop blocking the boy inside from coming out and being a rambling boy,[8] with a touch of madness.

[12:57 pm: 2hr, 55 min session]

[7] See "Kathy's Song" by Paul Simon[E2].
[8] See "Ramblin' Boy" by Tom Paxton[E3].

Epilogue

Modified from 2021-12-16 journal entry.

Thoughts for epilogue:

· Thank you reader, whether you have just skimmed over my ramblings or shared my epiphanies.
· The onus is on me now to follow through on the undisciplined creative burst you have just been subjected to.
· I will try.
· But I still don't know what style to adopt.

To Be Free

To be free, or not to be free: that is the question.
Whether 'tis nobler in the mind to suffer the lure and muddle of loose neuron firings,
Or to take reason against a sea of madness,
And by opposing end it?

Would that we could build a ship and travel together on a sea of madness to the shores of reason and find "a smoother pebble or a prettier shell than ordinary."

Addendum

Modified from 2021-12-22 journal entry.

Chloe is on the radiator underneath the window by my chair. She is licking her paws.

A lot of life and thought swirl going on. Not sure how to focus. Maybe I'll just write some scattered thoughts

Kimberly sent many emails last night, following up on discussion in the "infinite hamster regress" thread that I started when sending out my "Infinite Regress, Column 1" draft. Intriguingly, in her 12:53 am email she wrote, "Daxe has been instrumental in helping me formulate the insight I am currently finalizing, which I will share shortly."

So an interesting dynamic is arising.

As I am learning is usually the case, one of my first impulses is to react internally with my inner hurt/wary child voice, which I will let speak now.

I don't know how to sort through all of this.
I don't know how to play this game.
Why do I have to adapt?
And how would I?
They seem to be having fun.
But something seems wrong.
Why does something always seem wrong?

No, this inner voice is not me at all. This is my adult-self hijacking my genuine child inside, projecting all of his adult baggage on it.

I'm not really that upset and wounded. It's not my fault that my adult self is mourning over the fact that his younger, but still adult, self somehow went awry. He's trying to get me to brood and have that "a ha" moment.

In response to "Kimberly putting things in front of me," he made me say "But, but, but ... you can't just throw out your raw musings ..."[1]*.*

But now, I'm going to turn the tables on him. I say to my adult self[2]*:*

> *But, but, but ... you can't just throw out your raw musings [thoughts and feelings] and expect someone else [me!] to have the epiphany and do the hard [soul searching] follow through and do it on your boat.*

What an asshole, or maybe a whole ass, you are, adult self. You keep badgering me, asking "Why didn't you learn to be happy?"

Well, fuck you. First off, what right do have to mess with my mind.

I was happy enough before you started attacking me with questions in hindsight.

And on top of all that, you are going public and making everyone else think I am some morose little boy.

[1]See "but, but, but" on page 13, at the start of Part III: Try Again and page 21, in the middle of Part IV: Explanation

[2]This is a the same as what I (the inner child) said on page 13, but with bracketed word changes.

Jeez, I took things in stride. I found things I'm interested in that I could pursue, and I did.

And now you come along as some G-damn critic.

And you are still controlling my voice. You're making me talk about myself in the past tense, using extra layers of twisted-ness.

G-damn-it. I'm just a kid. Sure, I'm lonely sometimes. But mostly I find things to occupy myself with. And when there's people to play with, I do.

I have my own father to deal with, whatever is good and bad about that. Don't confuse me by offering to be a different and controlling father figure. You have your own kids to deal with anyway. They have their own issues.

Who knows, someday your kids may invent their own inner child voices to deal with the ramifications of having you as a father. And their inner children will complain about their adult selves trying to control them and trying to become a replacement father figure.

And now your are starting to fantasize about becoming a new Abraham, a patriarch paving the way for generations of his descendents to revere the moment when he smashed the idol of their internal God voice.

I can tell, you're already wondering what to call your new religion that has some kind of weird infinite regresses of overcoming constraints imposed by nested Gods.

It's no fun being burdened with your thoughts. "Uno-theism", the word you just invented, keeps popping in my head. But that's your weird adult game.

Why do you make everything so complicated? It's simple. You've even quoted the rule in the words of our Uncle Joe:

"Have fun and don't hurt anyone."

> *Look, I want to live my own full life. But we both know that can't happen while I'm stuck inside of you.*
>
> *Listen to me:*
>
> *I had my turn. No, actually, I'm having my turn in the perpetual now, where your "then" is now for me.*
>
> *Get away, you're confusing me. I'm satisfied with my lot. And I don't want to constrain myself to only live in your distorted memory.*
>
> *My teacher says I'm an easy going boy. And I get good grades because I like to learn. And I don't torture myself trying to do things I can't. And you can't make me. Ha ha, you can't make me.*
>
> *You can't make me write another word. So there.*
>
> *Good bye.*

Adult self pours some coffee and goes to the bathroom. He would like to comment, but will see how much time there is, what with responsibilities and all.

[Stop at 8:45 am, approx. 1hr 35 min session]

9:03 am: Note written by adult self.

Perhaps I will make this journal entry the basis for an addendum to Infinite Regress, Column 1.

Intro to Addendum:

During a restless night, I read some emails of Kimberly and Daxe, which were following up on my sending out the draft of Infinite Regress, Column 1.

To change the topic in my head, which I thought might help me get back to sleep, I opened up the kindle app in bed so that I could read from the book Happier by Tal Ben-Shahar, Ph.D.[E1]

The first words I saw were "infinite regress"[3]. The words were right in the middle of the top line of the screen when I opened the app. The full line is[4]:

> Their [questioning children's] relentless probing follows the pattern of the "infinitely regressive why" – regardless of the answer to a question, the child persists with another "Why?"

In my writing above, my inner child turned the tables on the adult me to get him to shut-up. It's interesting that I wrote this shortly after reading Ben-Shahar's suggestion for how to get the questioning child to stop their infinite regress[5]. His suggestion is made in the lines immediately following the line quoted above:

> [O]ne question allows an adult to end the onslaught of "whys" without any feelings of guilt or inadequacy. This question is "Why do you want to be happy?"

I guess I'm obligated now to try to write some insightful closing lines that connect to the body of the column.

Let's try the following.

⋮

Perhaps letting me inner child and madman voices out is already helping me to make progress connecting to the world.

[3]Actually, my mind didn't process the "ly" at the end of "infinite" in the actual text.

[4]Ben-Shahar, Tal. Happier (p. 31). McGraw-Hill Education. Kindle Edition. Location 544.

[5]Ben-Shahar, Tal; Ben-Shahar, Tal. Happier (p. 31). McGraw-Hill Education. Kindle Edition. Location 546.

I now notice how all of my reformulating the creativity + discipline pair was rather solipsistic. I was just focused on an individual mind, my own mind. I did mention Kimberly as a muse and Daxe and Kimberly as providing ideas, but what I wrote seems more worried about understanding my internal process than embracing external connection.

I don't want to minimize the fact that there is indeed a balancing act to strike between strong internal and strong external voices, but – and here is a new mini-epiphany – this addendum, and this whole essay in fact, would not have been possible without external input.

On page xi (kindle location 112) of "Happier", Ben-Shahar discusses a quote by Supreme Court Justice Oliver Wendall Holmes that says that a superficial simplicity that comes before considering complexity is far less meaningful than a simplicity that comes after struggling through complexity. I aspire to give my final words in this spirit. They are the same words we tell Chloe to release her after a training session:

Go play.

Description of Cover Picture

Written 2022-02-09.

I started with a picture of me and my wife, Denise Freed, in a lava tunnel in Iceland in February, 2020.

I added a series of shrinking copies of our picture with the following matlab code:

```
function res = infinite_regress_cover

res.imgFile = '2020-02-19_lava_tunnel.jpeg';
res.imgC = {};
res.subC = {};
res.img00   = imread(res.imgFile);
res.img0    = res.img00(1:(0.7*end), :,:);
res.imgC{1} = res.img0;

for k=2:6
  [res.imgC{k} res.subC{k}] = do(res.imgC{k-1},res.img0,k-1);
end

res.imgFinal = res.imgC{end};

close all hidden; imshow(res.imgFinal);
imwrite(res.imgFinal, [mfilename '.png']);

function [newimg sub] = do(img,img0, k)

 shiftVec = [0 0.3 0.6 0.7 0.8 0.9];
 shift    = shiftVec(k);
```

```
sub.img      = img0(1:2^k:end, 1:2^k:end,:);
[sub.Nx,sub.Ny,sub.Nz] = size(sub.img);

newimg = img;

[Nx Ny Nz] = size(newimg);

Ix = floor(0.05*Nx/2)+(1:sub.Nx);
Iy = floor((0.4+shift)*Ny/2)+(1:sub.Ny);

lx         = length(Ix);
ly         = length(Iy);
sub.interp = ones(lx, ly,3);
for i=1:lx
    for j=1:ly
        cx = abs((i - lx/2)/(lx/2)).^2;
        cy = abs((j - ly/2)/(ly/2)).^2;

        sub.interp(i,j,:) = 1 -max(cx,cy);
    end
end
newimg(Ix, Iy,:) = uint8(...
    sub.interp.*double(sub.img) + ...
    (1-sub.interp).*double(img(Ix,Iy,:)));
```

I included the png file output by the above matlab code, together with a vertically flipped image as a background on the cover page using the following LaTeX code:

```
\newcommand\BackgroundPic{
\setlength{\unitlength}{1in}
\put(0,0){
\parbox[b][\paperheight]{\paperwidth}{
\vfill
\begin{center}
\baselineskip=0pt
```

```
\includegraphics[
  height=\paperheight*2/5,
  trim={2in 0in 0in 0in},
  keepaspectratio]{./infinite_regress_cover.png}\\
\scalebox{-1}[1]{\includegraphics[
  height=\paperheight*2/5,
  trim={2in 0in 0in 0in},
  angle=180,
  keepaspectratio]{./infinite_regress_cover.png}}
\end{center}
\vfill
}}}
```

End Notes (html links)

Chapter 0

1. the inaugural issues of the journal "Tandeta", entitled "Epiphany", published by Tandeta Books
https://www.tandetabooks.com

2. "Infinite Placeholder Regress" by Scott Axelrod in (Tandeta Journal 2022.01: "Epiphany"
https://www.tandetabooks.com/tandeta-journal-january-2022-infinite-placeholder-regress/

Chapter 1

1. "Infinite Regress - Virtue of Vice?"
https://www.fil.lu.se/hommageawlodek/site/papper/MaurinAnnaSofia.pdf

2. Hommage à Wlodek. Philosophical Papers Dedicated to Wlodek Rabinowicz. Ed. T. Rønnow-Rasmussen, B. Petersson, J. Josefsson & D. Egonsson, 2007.
www.fil.lu.se/hommageawlodek

3. "We're Not Gonna Take It" by the Who
https://www.azlyrics.com/lyrics/who/werenotgonnatakeit.html

4. Off the Deep End: Diary of a Mathematician by Scott Axelrod
https://www.amazon.com/Off-Deep-End-Diary-Mathematician/dp/0988243806

Chapter 2

1. Wikipedia entry on Paul-Michel Foucault
https://en.wikipedia.org/wiki/Michel_Foucault

2. Madness and Civilization: A History of Insanity in the Age of Reason by Paul-Michel Foucault
https://en.wikipedia.org/wiki/Madness_and_Civilization

3. Tandeta Books
https://www.tandetabooks.com

4. I hate You – Don't Leave Me: Understanding the Borderline Personality by Jerold J. Kreisman
https://www.psychologytoday.com/us/blog/i-hate-you-dont-leave-me/202109/what-is-new-and-what-isnt-in-our-understanding-bpd

Chapter 3

1. interview of Isadore Singer with Science Lives
https://www.youtube.com/watch?v=XYsLQsrSQLc

2. Isaac Newton's famous quote
https://www.brainyquote.com/quotes/isaac_newton_387031

Chapter 4

1. Speakable and Unspeakable in Quantum Mechanics by J.S. Bell
https://www.cambridge.org/core/books/speakable-and-unspeakable-in-quantum-mechanics/E0D032E7E7EDEF4E4AD09F458F2D9DB7

2. "Links Between Physics and Set Theory" by Bruno W. Augensten, Chaos, Solitons, and Fractals, Vol 7, Issue 11, Nov. 1996, pp. 1761-1798]
https://www.sciencedirect.com/science/article/abs/pii/S0960077996000550

Chapter 5

1. "Black Holes: the Harmonic Oscillators of the 21st Century", talk by Andy Strominger in Perimeter Institute Colloquium, March 2014]
https://pirsa.org/14030104/

2. "Kathy's Song" by Paul Simon
https://www.azlyrics.com/lyrics/simongarfunkel/kathyssong.html

3. "Ramblin' Boy" by Tom Paxton
https://www.azlyrics.com/lyrics/tompaxton/ramblinboy.html

Chapter 7

1. Happier by Tal Ben-Shahar, Ph.D.
https://uvolution.net/happier-book-summary/

Made in the USA
Monee, IL
04 November 2026

67386541R00026